GCSE Maths Challeng

Tough problems with solut

ISBN 9781520166438

Introduction

This booklet contains a selection of mathematics problems together with their solutions. The problems are intended to be challenging for students who are preparing for GCSE mathematics examinations and are expecting to achieve a grade 8 or 9. Some of the questions could conceivably appear on a GCSE exam, others are significantly more difficult.

Questions appear one per page and are followed immediately by their solutions. This way you will not accidentally see the solution while you are still thinking about a question but you will find it convenient to check your solution when you are ready. The solutions given are not intended to be as they would be written as a response to an examination question, rather they are intended to be sufficiently detailed for the reasoning process to be clear, given some effort on the part of the reader.

The author is a qualified teacher of mathematics and a professional mathematics tutor with over 20 years' experience teaching mathematics to small groups and individuals.

Copyright © 2016 Mark Ritchings

Question 1

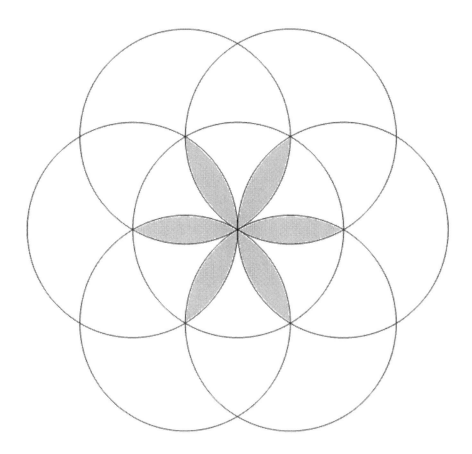

The area of each circle is 4π cm².

What area is shaded?

Solution 1

$4\pi = \pi r^2 \Rightarrow r = 2.$

Each of the six parts of the shaded area is made up of two segments.

segment area = sector area - triangle area.

$= \dfrac{4\pi}{6} - \dfrac{1}{2} \times 4 \sin 60°$

$= \dfrac{4\pi}{6} - \sqrt{3}$

The shaded area is $12\left(\dfrac{4\pi}{6} - \sqrt{3}\right) = (8\pi - 12\sqrt{3})$cm².

Question 2

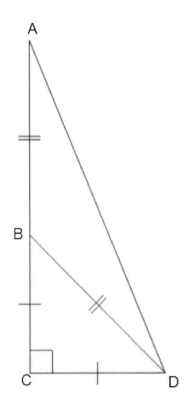

$BC = 1$.
Find the exact value of $\tan 22.5°$.

Solution 2

Angle $CBD = 45°$.
Angle $ABD = 135°$.
Angle $BAD = \frac{180°-135°}{2} = 22.5°$.
$AB = BD = \sqrt{2}$.
$AC = 1 + \sqrt{2}$.
$\tan 22.5° = \frac{CD}{AC} = \frac{1}{1+\sqrt{2}} = -1 + \sqrt{2}$.

Question 3

If the density of gold is 19.3 tonnes/m^3 and the price of gold is £32/g, find the value of a cube of gold the edges of which are 5 cm long.
You may use a calculator.

Solution 3

1 tonne = 1000 kg.
1 kg = 1000 g.
1 tonne = 1,000,000 g.
1 m^3 = (1 m)^3 = (100 cm)^3 = 1,000,000 cm^3.
19.3 tonnes/m^3 = 19,300,000 g/1,000,000 cm^3 = 19.3 g/cm^3.
The volume of the cube is 5 x 5 x 5 cm^3 = 125 cm^3.
The mass of the cube is 125 cm^3 x 19.3 g/cm^3 = 2412.5 g.
The value of the cube is 2412.5 g x £32/g = £77,200.

Question 4

The first three terms of a quadratic sequence are 3, 6 and 13.
Which term of the sequence is equal to 409?

Solution 4

If the nth term of the sequence is $an^2 + bn + c$ then the first term (when $n = 1$) is $a + b + c$, therefore $a + b + c = 3$.
Similarly the second term is $a \times 2^2 + b \times 2 + c = 4a + 2b + c$ and so $4a + 2b + c = 6$.
A third equation can be found in the same way to give three equations with three unknowns.

You can write the first and second differences in a table like this.

$a + b + c = 3$ $\qquad\qquad$ $4a + 2b + c = 6$ $\qquad\qquad$ $9a + 3b + c = 13$
$\qquad\qquad 3a + b = 3 \qquad\qquad\qquad 5a + b = 7$
$\qquad\qquad\qquad\qquad 2a = 4$

$a = 2$.
$b = 3 - 6 = -3$.
$c = 3 - 2 + 3 = 4$.

The nth term of the sequence is $2n^2 - 3n + 4$.

One of the values of n for which $2n^2 - 3n + 4 = 409$ is the term number required.

$2n^2 - 3n - 405 = 0 \quad \Rightarrow \quad n = \dfrac{3 + \sqrt{3^2 - 4 \times 2 \times -405}}{2 \times 2}$

$= \dfrac{3 + \sqrt{3249}}{4} = \dfrac{3 + 57}{4} = 15$.

409 is the 15th term of the sequence.

Question 5

A, B and P are points on a circle, centre O. Q is the midpoint of AB.
POQ is a straight line. AB = 14 cm. PQ = 49 cm.
Calculate the radius of the circle.

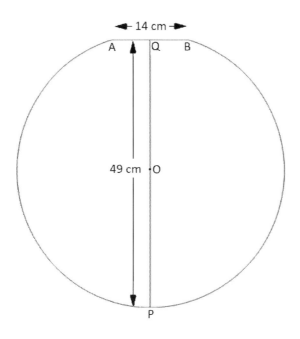

Solution 5

$QB = 7$ cm.
$OB = OP = r$, the radius of the circle.
$OQ = 49 - OP = 49 - r$.
By Pythagoras' theorem:
$OB^2 = OQ^2 + QB^2$
$r^2 = (49 - r)^2 + 7^2$
$r^2 = 2401 - 98r + r^2 + 49$
$98r = 2450$
$r = \dfrac{2450}{98}$
$r = 25$ cm.

Question 6

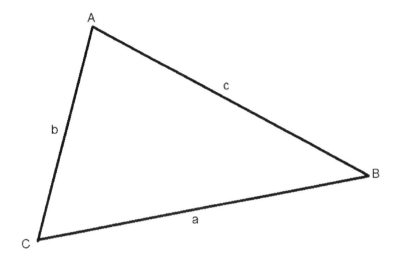

Show that the area of the triangle is $\frac{1}{2}ab \sin C$.

Solution 6

Let the perpendicular distance from A to BC be h.

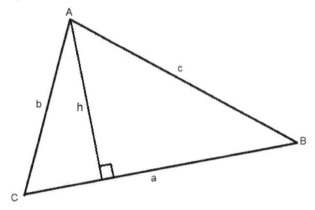

$\sin C = \dfrac{h}{b}$

$h = b \sin C$

The area of the triangle is $\dfrac{a \times h}{2} = \dfrac{1}{2} ab \sin C$.

Question 7

The diagram shows an equilateral triangle and a sector of a circle the centre of which is one of the vertices of the triangle. Find the area shaded in the diagram, giving your answer in its exact form.

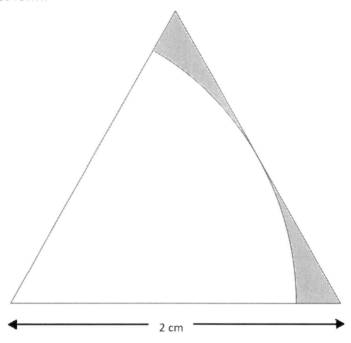

2 cm

Solution 7

The area of the triangle is found using the formula
$Area = \frac{1}{2} ab \sin C$.

The area of the triangle is $\frac{1}{2} \times 2 \times 2 \times \sin 60 = \sqrt{3}$.

The radius of the circle is the height of the triangle.
This can be found by considering the area of the triangle as $\frac{base \times height}{2}$

This gives the equation $\frac{2h}{2} = \sqrt{3}$. Therefore the radius of the circle is $\sqrt{3}$.

The area of the sector is $\frac{\pi r^2}{6} = \frac{\pi \times 3}{6} = \frac{\pi}{2}$.

The shaded area is therefore $\left(\sqrt{3} - \frac{\pi}{2}\right) cm^2$.

Question 8

Find c in terms of a, b and C.

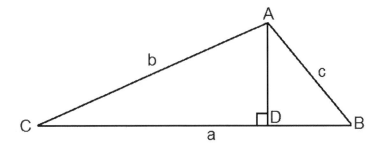

Solution 8

Let $CD = x$ and let $AD = h$.
Then $\cos C = \frac{x}{b} \Rightarrow x = b \cos C$ and by Pythagoras' theorem
$h^2 = b^2 - x^2 = c^2 - (a-x)^2$.
Expanding and simplifying gives
$b^2 - x^2 = c^2 - (a^2 - 2ax + x^2)$.
Removing the brackets $b^2 - x^2 = c^2 - a^2 + 2ax - x^2$.
Adding x^2 to both sides, $b^2 = c^2 - a^2 + 2ax$
Substituting $b \cos C$ for x,
$b^2 = c^2 - a^2 + 2ab \cos C$
Rearranging a little,
$c^2 = a^2 + b^2 - 2ab \cos C$
$c = \sqrt{a^2 + b^2 - 2ab \cos C}$

Question 9

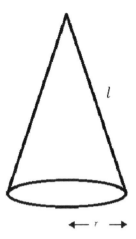

If the slant height of a cone is l and its radius is r, show that the curved surface area is $\pi r l$.

Solution 9

The circumference of the base of the cone is $2\pi r$. This is the arc length of the corresponding sector that the cone could be formed from.

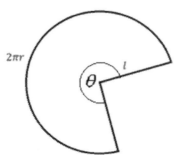

If the sector angle is θ then the arc length is $\frac{\theta}{360} \times 2\pi l$.

$\frac{\theta}{360} \times 2\pi l = 2\pi r$

Dividing both sides of this equation by 2π we have

$\frac{\theta}{360} \times l = r \Rightarrow \theta = \frac{360 r}{l}$

The curved surface area of the cone is the sector area, $\frac{\theta}{360} \times \pi l^2$.

Substituting for θ,

$Area = \frac{\frac{360r}{l}}{360} \times \pi l^2$

$= \frac{\pi r l^2}{l}$

$= \pi r l.$

Question 10

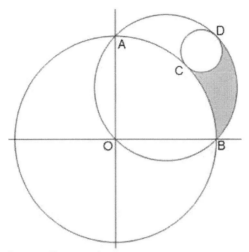

The diagram shows three circles.
O is the centre of the largest circle.
AB is the diameter of one of the circles.
C and D lie on bisector of angle AOB.
The area of the largest circle is π.
Find the exact area of the shaded region in its simplest form.

Solution 10

Let the circles, in order of size from smallest to largest be C_1, C_2 and C_3 and let their areas be A_1, A_2 and A_3.

$$A_3 = \pi r^2 \Rightarrow r^2 = \frac{A}{\pi} \Rightarrow r = \sqrt{\frac{A}{\pi}}$$

The radius of C_3 is therefore 1.
By Pythagoras' theorem $AB^2 = OA^2 + OB^2 = 2$
$AB = \sqrt{2}$. The radius of C_2 is $\frac{\sqrt{2}}{2}$.
$OD = AB = \sqrt{2}$ and $OC = 1$ therefore $CD = \sqrt{2} - 1$ and the radius of C_1 is $\frac{\sqrt{2}-1}{2}$.

The area of segment ABC is $\frac{A_3}{4} - \frac{OA \times OB}{2} = \frac{\pi}{4} - \frac{1}{2} = \frac{\pi-2}{4}$.

The required area is half of $\frac{A_2}{2} - A_1 - \frac{\pi-2}{4}$

$$= \frac{\frac{A_2}{2} - A_3 - \frac{\pi-2}{4}}{2}$$

$$= \frac{2A_2 - 4A_3 - \pi + 2}{8}$$

$$= \frac{2\pi\left(\frac{\sqrt{2}}{2}\right)^2 - 4\pi\left(\frac{\sqrt{2}-1}{2}\right)^2 - \pi + 2}{8}$$

$$= \frac{2 - (3 - 2\sqrt{2})\pi}{8}$$

Question 11

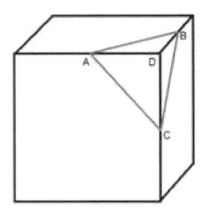

The diagram shows a cube. A, B and C are midpoints of edges. The volume of the cube is 64cm³.
The corner indicated in the diagram is removed from the cube.
What is the surface area of the smaller piece?

Solution 11

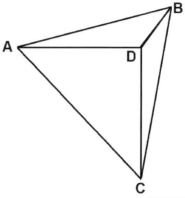

The length of each edge of the cube is $\sqrt[3]{64 cm^3} = 4\ cm$.
In triangle ADC the sides AD and DC are 2cm and angle ADC is a right angle. The total area of the triangles ADB, BCD and ACD is $3 \times \frac{2 \times 2}{2} = 6\ cm^2$.
$AC^2 = AD^2 + DC^2 = 2^2 + 2^2 = 8$.
$AC = AB = BC = \sqrt{8} = 2\sqrt{2}$. Triangle ABC is equilateral.
The area of triangle ABC is $\frac{1}{2} \times 2\sqrt{2} \times 2\sqrt{2} \times \sin 60 = 2\sqrt{3}$.
The total surface area of the corner piece is $(6 + 2\sqrt{3})\ cm^2$.

Question 12

There are 12 discs in a bag. The majority of the discs are red and the rest are blue. Two discs are taken, at random, from the bag. The first disc is not replaced in the bag. The probability that one disc of each colour is taken is $\frac{9}{22}$. Find the number of red discs and the number of blue discs.

Solution 12

Let the number of red discs be n. The number of blue discs is then $12 - n$.

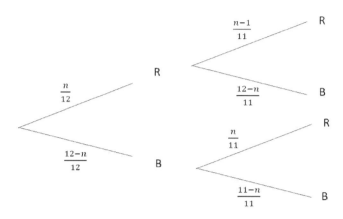

The probability of taking one disc of each colour is
$$\frac{n}{12} \times \frac{12-n}{11} + \frac{12-n}{12} \times \frac{n}{11} \text{ so } 2 \times \frac{n}{12} \times \frac{12-n}{11} = \frac{9}{22}$$
Multiplying both sides of the equation by 132 gives
$2n(12 - n) = 54$
$n(12 - n) = 27$
$12n - n^2 = 27$
$n^2 - 12n + 27 = 0$
$(n - 3)(n - 9) = 0$
$n = 9$ or $n = 3$, but "the majority of the discs are red" so $n = 9$.
There are 9 red discs and 3 blue discs.

Question 13

The diagram shows a solid glass cylinder.

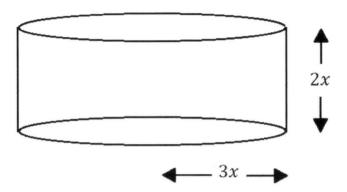

The cylinder has a base radius of $3x$ and a height of $2x$.
The cylinder is melted down and made into a hemisphere of radius r.
No glass is lost in the process.
Find an expression for r in terms of x.

Solution 13

The volume, V, of a cylinder of radius r and height h is given by $V = \pi r^2 h$ so the volume of the glass cylinder in terms of x is $V = \pi (3x)^2 \times 2x = 9 \times 2 \times \pi x^3$.

The volume, V, of a sphere of radius r is given by $V = \frac{4}{3}\pi r^3$ so the volume of the hemisphere in terms of r is $\frac{2}{3}\pi r^3$.

The volume of the cylinder is equal to the volume of the hemisphere. This gives the equation $9 \times 2 \times \pi x^3 = \frac{2}{3}\pi r^3$.

Multiplying both sides of the equation by 3 gives $27 \times 2\pi x^3 = 2\pi r^3$.

Dividing both sides of the equation by 2π gives $27x^3 = r^3$.

And so $r = \sqrt[3]{27x^3}$.

$r = 3x$.

Question 14

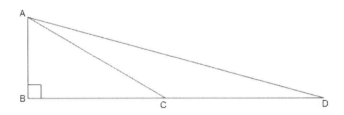

$AB = 1$. $AC = CD = 2$.
Find the exact value of $\tan 75°$.
Find the exact value of $\tan 15°$.

Solution 14

$BC^2 = AC^2 - AB^2 = 3.$
$BC = \sqrt{3}.$
$\sin ACB = \frac{1}{2}.$
Angle $ACB = 30°.$
Angle $ACD = 150°.$
Angle $CAD = \frac{180° - 150°}{2} = 15°.$
Angle $BAD = 75°.$
$\tan 75° = 2 + \sqrt{3}.$
$\tan 15° = \frac{AB}{BD} = \frac{1}{2+\sqrt{3}} = 2 - \sqrt{3}.$

Question 15

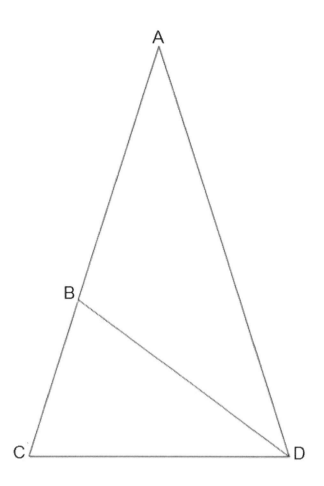

$BD = CD$.
Show that $\sin ABD = \sin CBD$.

Solution 15

Angle CBD = angle BCD.

$$\frac{\sin(BCD)}{AD} = \frac{\sin(BAD)}{CD} \qquad \text{(Sine rule, triangle ACD)}$$

$$\sin(BCD) = AD\,\frac{\sin(BAD)}{CD}$$

$$\frac{\sin(ABD)}{AD} = \frac{\sin(BAD)}{BD} \qquad \text{(Sine rule, triangle ABD)}$$

$$\sin(ABD) = AD\,\frac{\sin(BAD)}{BD} = AD\,\frac{\sin(BAD)}{CD}.$$

$$\sin(ABD) = \sin(BCD) = \sin(CBD).$$

Question 16

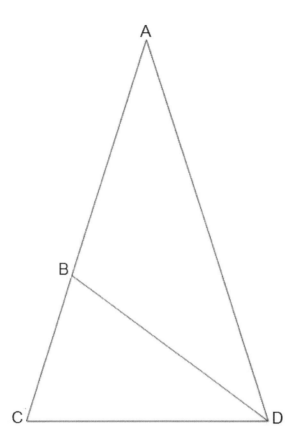

$AB = BD = DC$.
$AC = AD$.
Following on from question 15, find the exact value of $\sin 18°$.

Solution 16

Let angle $BAD = x$.
Angle $ADB = x$.
Angle $CBD = $ angle $BCD = 2x$.
$5x = 180$.
$x = 36°$.
Let $AB = 1$ and $BC = y$.

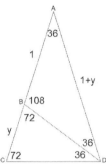

$\sin 72° = \sin 108°$ \qquad (See question 15)

$\dfrac{\sin 108°}{1+y} = \sin 36° \Rightarrow \sin 108° = (1+y) \sin 36°$

$\dfrac{\sin 36°}{y} = \sin 72°$

$(1+y) \sin 36° = \dfrac{\sin 36°}{y}$

$1 + y = \dfrac{1}{y} \Rightarrow y^2 + y - 1 = 0$

$y = \dfrac{-1+\sqrt{5}}{2}$ and $\sin 18° = \dfrac{-1+\sqrt{5}}{4}$

Question 17

The surface area of this regular tetrahedron is $\sqrt{3}$.

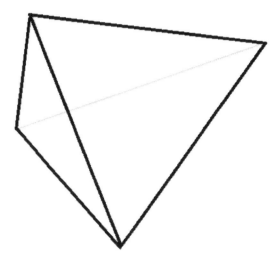

What is the volume of the tetrahedron?

Solution 17

The diagram below shows the net of the tetrahedron.

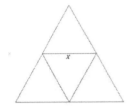

Let the length of each edge be x.
The area of each face is then
$$\frac{1}{2}x^2 \sin 60 = \frac{1}{2}x^2 \frac{\sqrt{3}}{2} = \frac{\sqrt{3}}{4}x^2.$$

The area of all four triangles is therefore $\sqrt{3}x^2$.
The surface area of the tetrahedron is $\sqrt{3}$ and so $x = 1$.
The diagram below shows the base of the tetrahedron.

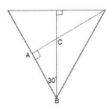

$AB = \frac{1}{2}$.

$\cos 30 = \frac{AB}{BC} \Rightarrow$

$BC = \frac{AB}{\sqrt{3}/2} = \frac{1}{\sqrt{3}}$.

Let V be the vertex directly above C.

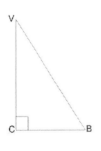

The height of the tetrahedron is then
$\sqrt{VB^2 - BC^2} = \sqrt{1 - \frac{1}{3}}$ and the volume of the tetrahedron is $\frac{\text{base area} \times \text{height}}{3}$

$= \frac{1}{3} \times \frac{\sqrt{3}}{4} \times \sqrt{\frac{2}{3}}$

$= \frac{\sqrt{2}}{12}$.

Question 18

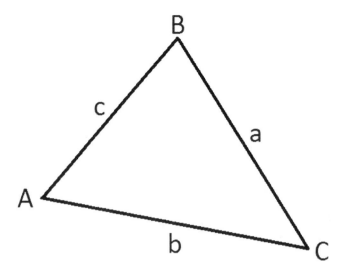

Find a formula for the area of triangle ABC, in terms of A, b and C.

Solution 18

By the sine rule $\dfrac{c}{\sin C} = \dfrac{b}{\sin B}$ so

$$c = \dfrac{b \sin C}{\sin B} = \dfrac{b \sin C}{\sin(A+C)}.$$

The area of the triangle is $\dfrac{1}{2} bc \sin A$.

Substituting for c gives $\text{Area} = \dfrac{b^2 \sin A \sin C}{2 \sin(A+C)}.$

Question 19

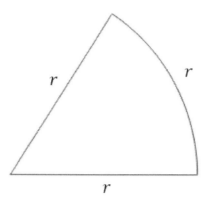

A sector of a circle has radius r cm, arc length r cm and area r cm².
Find the value of r.

Solution 19

The arc length is $\frac{\theta}{360} \times \pi \times d$ so $\frac{\theta}{360} \times \pi \times 2r = r$.

It follows that $\theta = \frac{180}{\pi}$.

The area of the sector is $\frac{\theta}{360} \times \pi \times r^2 = r$.

Substituting in the value of θ we have $\frac{\frac{180}{\pi}}{360} \times \pi \times r^2 = r$.

Simplifying this gives $\frac{r^2}{2} = r$.

This leads to $r^2 - 2r = 0$.

Factorising gives $r(r-2) = 0$ and so $r = 2$.

Question 20

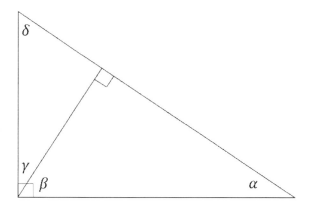

Show that the three triangles above are similar.

Solution 20

$\alpha + \beta = 90°$, $\beta + \gamma = 90°$ and $\gamma + \delta = 90°$ therefore $\alpha = \gamma$ and $\beta = \delta$. The three triangles have the same three angles. The triangles are therefore similar.

Question 21

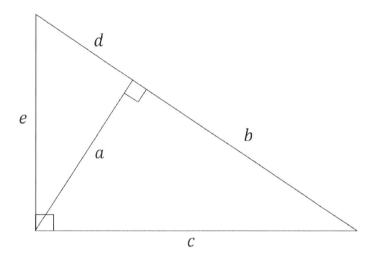

The triangles above are similar as shown in question 19.
Observe that $\frac{b}{a} = \frac{a}{d}$ and that $\frac{c}{b} = \frac{b+d}{c}$.

Prove that $a^2 + b^2 = c^2$.

Solution 21

$\dfrac{b}{a} = \dfrac{a}{d} \Rightarrow bd = a^2.$

$\dfrac{c}{b} = \dfrac{b+d}{c} \Rightarrow c^2 = b^2 + bd.$

Hence $a^2 + b^2 = c^2.$

Question 22

A car has wheels with a radius of 25cm. If the car travels at 100km/h how many revolutions does a wheel make in one second?

Solution 22

The circumference of wheel is $2\pi \times 25 = 50\pi$ cm.
100km=100x1000x100=10000000cm. In travelling 100km the wheel must turn through $\frac{10000000}{50\pi}$ revolutions. At a speed of 100km/h that is $\frac{200000}{\pi}$ revolutions in one hour. In one second the number of revolutions is $\frac{200000}{3600\pi}$. That's about 17.7 revolutions per second.

Question 23

An equilateral triangle is inscribed in a circle of radius 2cm. What is the area of the triangle?

A square is inscribed in a circle of radius 5cm. What is the area of the square?

An n sided regular polygon is inscribed in a circle of radius r. What is the area of the polygon?

What is the area of a polygon with one million sides, inscribed in a circle of radius 1cm?

Solution 23

Dividing the n sided polygon into n congruent triangles with a vertex at the centre of the circle and bisecting each triangle gives a right-angled triangle with an angle of $\frac{360}{n}$ degrees at the centre of the circle. The hypotenuse of the triangle is the radius of the circle. The area of each triangle is $\frac{1}{2}r^2 \sin\frac{360}{n}$.
The area of the polygon is therefore $\frac{1}{2}nr^2 \sin\frac{360}{n}$.

For a triangle and a circle of radius 2cm the area is $\frac{1}{2} \times 3 \times 2^2 \times \sin 120 = 3\sqrt{3}$ cm².

For a square and a circle of radius 5cm the area is $\frac{1}{2} \times 4 \times 5^2 \times \sin 90 = 50$cm². There is an easier way to get this answer.

If the polygon has one million sides and the radius of the circle is 1cm then the area is $\frac{1}{2} \times 1000000 \times \sin\frac{360}{1000000}$ cm² and that is very close to π cm², as you would expect.

Question 24

If the sides of the black equilateral triangle are 10cm and the other straight lines are all 1.5cm from the sides of the triangle. What is the shaded area?

Solution 24

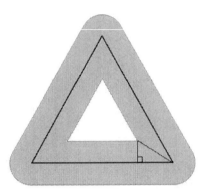

By considering the right-angled triangle shown above the length of the sides of the smaller equilateral triangle can be seen to be $10 - 2 \times \frac{1.5}{\tan 30°} = 10 - 3\sqrt{3}$.

The required area is then

$\frac{1}{2} \times 10^2 \times \frac{\sqrt{3}}{2} - \frac{1}{2}(10 - 3\sqrt{3})^2 \times \frac{\sqrt{3}}{2} + 3 \times 10 \times 1.5 + \pi \times 1.5^2$

$= 25\sqrt{3} - \frac{1}{2}(127 - 60\sqrt{3}) \times \frac{\sqrt{3}}{2} + 45 + \frac{9\pi}{4}$

$= 25\sqrt{3} - \frac{127\sqrt{3}}{4} + 45 + 45 + \frac{9\pi}{4}$

$= \left(90 - \frac{27\sqrt{3}}{4} + \frac{9\pi}{4}\right) cm^2.$

Question 25

Make t the subject of the formula $s = ut + \frac{1}{2}at^2$.

Solution 25

$$s = ut + \frac{1}{2}at^2$$

$$at^2 + 2ut - 2s = 0$$

$$t = \frac{-2u \pm \sqrt{(2u)^2 - 4a(-2s)}}{2a}$$

$$t = \frac{-u \pm \sqrt{u^2 + 2as}}{a}$$

Question 26

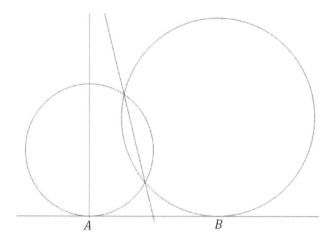

The diagram shows circles with equations

$x^2 + (y - a)^2 = a^2$ and $(x - b)^2 + (y - c)^2 = c^2$

together with the straight line that passes through their points of intersection.

Find the coordinates of the point where this lines intersects the x axis.

Hence show that the line bisects AB.

Solution 26

Expanding the brackets in both equations gives

$x^2 + y^2 - 2ay + a^2 = a^2$

$x^2 - 2bx + b^2 + y^2 - 2cy + y^2 = y^2$

Simplifying gives

$x^2 + y^2 = 2ay$

$x^2 + y^2 = 2bx + 2cy - b^2$

It follows that $2ay = 2bx + 2cy - b^2$.

This is the equation of the straight line. Where the line crosses the x axis $y = 0$ and so

$2bx - b^2 = 0$

$b(2x - b) = 0$

$b \neq 0$ therefore $x = \frac{b}{2}$ at the point of intersection and the required coordinates are $\left(\frac{b}{2}, 0\right)$.

A is the point $(0,0)$ and B is the point $(b, 0)$. The midpoint of AB is $\left(\frac{b}{2}, 0\right)$ so it is clear that the line bisects AB.

Question 27

Consider the equations $x^2 = 26$ and $(5 + e)^2 = 26$. By expanding the brackets in the second equation find an estimate for the square root of 26.

Hint: If e is close to zero then e^2 is very close to zero.

———

Estimate the square root of 34.

———

Given $\sqrt{n}$ is close to m, where m is an integer, find a formula that gives a better approximation to $\sqrt{n}$.

Solution 27

$(5 + e)^2 = 26$

$25 + 10e + e^2 = 26$

Ignoring the term e^2, as it is relatively small, you can see that

$e \approx 0.1$

The square root of 26 is approximately 5.1.

$(6 - e)^2 = 34$

$36 - 12e + e^2 = 34$

$12e \approx 2$

$e \approx \frac{1}{6} = 0.1\dot{6}$

$\sqrt{34} \approx 5.83$

Let the square root of n be $m + e$. Then $(m + e)^2 = n$.

$m^2 + 2me + e^2 = n$

$e \approx \frac{n - m^2}{2m}$

$\sqrt{n} \approx m + \frac{n - m^2}{2m}$

Question 28

The curve with equation $y = x^2 + bx + c$ passes through the points $(1,3)$ and $(4,9)$. Find the value of b and the value of c.

Solution 28

Substituting $x = 1$ and $y = 3$ into the equation gives $3 = 1 + b + c$.

Substituting $x = 4$ and $y = 9$ into the equation gives $9 = 16 + 4b + c$.

This gives the simultaneous equations $b + c = 2$ and $4b + c = -7$.

It follows that $3b = -9$, $b = -3$ and $c = 5$.

Question 29

A city has three detectors that can determine how far away a threat is. One detector is located at the centre of the city. The other detectors are one kilometre north of the city and one kilometre east of the city.

The threat is 3.742km from the city centre, 3km from the detector in the north and 3.317km from the detector in the east.

Where is the threat?

Solution 29

Let the positions of the detectors be (0,0,0) for the city centre, (1,0,0) for the detector in the east and (0,1,0) for the detector in the north.

Let the position of the threat be (x, y, z).

By Pythagoras' theorem the squares of the distances are

$$x^2 + y^2 + z^2 = 3.742^2 \approx 14 \qquad (1)$$

$$(x-1)^2 + y^2 + z^2 = 3.317^2 \approx 11 \qquad (2)$$

$$x^2 + (y-1)^2 + z^2 = 9 \qquad (3)$$

Expanding, simplifying and subtracting equations gives

(1) − (2) $2x - 1 = 3$
(1) − (3) $2y - 1 = 5$

So $x \approx 2$, $y \approx 3$ and $z \approx \sqrt{14 - 2^2 - 3^2}$

Assuming that the threat is above ground its position is approximately (2,3,1).

Question 30

The frequency of a note played on a stringed instrument is inversely proportional to the length of the string. Plucking the A string of a guitar produces a note with a frequency of 110Hz. Playing an E at the seventh fret of this string produces a note with a frequency close to 165Hz. What is the ratio of the lengths of the parts of the string above and below the seventh fret?

Halving the length of the string doubles the frequency. Fretted instruments have the twelfth fret positioned below the middle of the strings. By what percentage does the gap between frets decrease from one fret to the next?

Using this percentage, what is the ratio found above?

Solution 30

The reciprocal of $\frac{165}{110}$ is $\frac{2}{3}$. The part of the string above the seventh fret is $\frac{2}{3}$ the length of the whole string. The ratio of the lengths of the parts is 2:1.

The solution of this part is given in more detail below.

$$F = \frac{K}{L} \Rightarrow K = FL$$

If the vibrating length of the string when playing A is L_A and the frequency is $110Hz$ then $K = 110L_A$.

$$L_E = \frac{K}{F_E} = \frac{110L_A}{165} = \frac{2}{3}L_A \text{ and the ratio is as found above.}$$

Assume that the length of the vibrating string for a note played at the third fret is r times the length of the vibrating string for a note played at the second fret and so on for all frets.

Then, if the open string has a length L, the string lengths for notes played at the first and second frets are Lr and Lr^2.

At the twelfth fret the length is $Lr^{12} = \frac{L}{2}$.

So $r^{12} = \frac{1}{2}$ and $r = \left(\frac{1}{2}\right)^{\frac{1}{12}} = 0.94387..$

The length is reduced by about 5.61% for each fret.

Using the value of r above, the length of the vibrating string at the seventh fret is Lr^7. That is about 66.74% of the full length of the string. The ratio of the parts of the string is then 66.74:33.26 or about 2:1.

Question 31

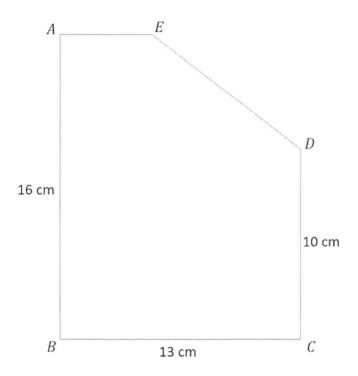

The diagram shows a pentagon $ABCDE$. The angles EAB, ABC and BCD are right angles. The lengths of AE and ED are in the ratio 1:2.

Calculate the area of the pentagon.

Solution 31

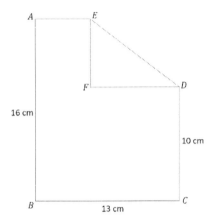

In the diagram above angle EFD is a right angle.

Let the length of AE be x cm and the length of ED be $2x$ cm.

In the right angled triangle EFD, $EF = 6$ cm, $FD = (13 - x)$ cm and $ED = 2x$ cm.

By Pythagoras' theorem $6^2 + (13 - x)^2 = (2x)^2$.

Expanding, simplifying and rearranging gives $3x^2 + 26x - 205 = 0$.

So $(3x + 41)(x - 5) = 0$

$AE = 5$ cm and $FD = 8$ cm.

The area is $10 \times 13 + 5 \times 6 + \frac{8 \times 6}{2} = 184$ cm^2.

Question 32

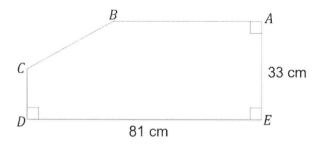

In the diagram above the ratio of the lengths of AB, BC and CD is 3:2:1.

Calculate the perimeter of the pentagon $ABCDE$.

Solution 32

As in solution 31, consider the right angled triangle, the time with BC as the hypotenuse and the sides AB, BC and CD have lengths $3x$, $2x$ and x.
The sides of the triangle are $(33 - x)$ cm, $(81 - 3x)$ cm and $2x$ cm.

$(33 - x)^2 + (81 - 3x)^2 = (2x)^2$

$6x^2 - 552x + 7650 = 0$

$x^2 - 92x + 1275 = 0$

$(x - 17)(x - 75) = 0$

$x = 17$

The perimeter is $81 + 33 + 6 \times 17 = 216$ cm

Printed in Germany
by Amazon Distribution
GmbH, Leipzig